Ayushman Bharat Digital Mission

From Infrastructure to Impact: A Tri-Lens Assessment of India's Digital Health System

Pooja Kadambi
Divya Ajitsaria
Peter Joseph

Ayushman Bharat Digital Mission

From Infrastructure to Impact: A Tri-Lens Assessment of India's Digital Health System

ELIVA PRESS

ISBN: 978-99993-4-602-3

Cover Design: Eliva Press
Cover Image: www.ingimage.com
Email: info@elivapress.com
Website: www.elivapress.com

This Eliva Press imprint is published by the registered company Eliva Press Global Ltd. part of Eliva Press S.R.L. Publishing Group

The registered company address is: Pope Hennessy Street Level 2, Hennessy Tower Port Louis, Mauritius
Eliva Press S.R.L. Publishing Group legal address is: Bd. Cuza-Voda 1/4 of. 21, Chisinau, Moldova, Europe

AYUSHMAN BHARAT DIGITAL MISSION

FROM INFRASTRUCTURE TO IMPACT: A TRI-LENS ASSESSMENT OF INDIA'S DIGITAL HEALTH SYSTEM

Author
Pooja Kadambi, Co-founder Just Yukti

Author
Divya Ajitsaria , Co-founder Just Yukti

Author
Dr. Peter T Joseph, Founder, Pathros Labs

Author
Yajur Arora

Executive Summary

India's Ayushman Bharat Digital Mission (ABDM) is among the world's most ambitious national digital-health programmes, implementing a federated, standards-based architecture for secure, consented health-data exchange across a mixed health system (NHP 2017; NDHB 2019). Launched in 2021 under the National Health Authority (NHA), ABDM operationalises the policy intent that "data should follow the patient" through core building blocks: ABHA digital IDs, Health Facility Registry (HFR), Health Professional Registry (HPR), Personal Health Records (PHR), and the National Health Claims Exchange (NHCX) (NHA Annual Report 2022–23; ABDM Building Blocks; NRCeS FHIR Implementation Guide 2024).

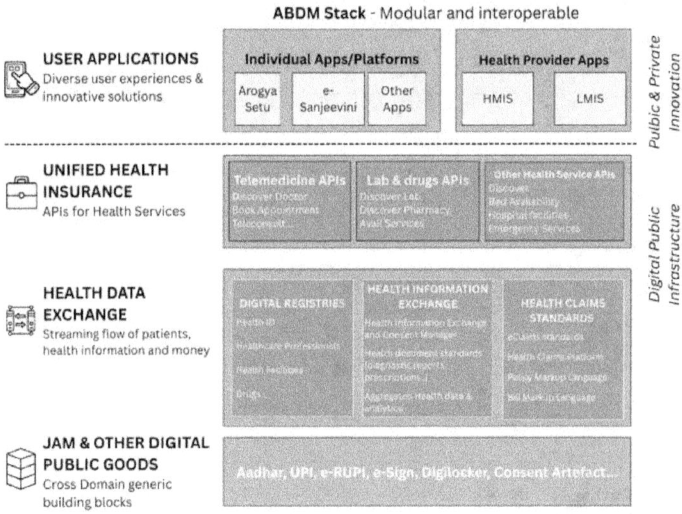

Fig 1. The ABDM Stack

This whitepaper provides a tri-lens assessment—policy/benchmark review (WHO 2020–25; OECD 2022/23), health-economic evaluation (using ICER/NPV/BCR/ SROI models calibrated to an India-specific affordability threshold, λ_IN), and stakeholder sentiment analysis (N=20 interviews). Together, these lenses assess whether India's digital health infrastructure is translating into equitable, measurable public value.

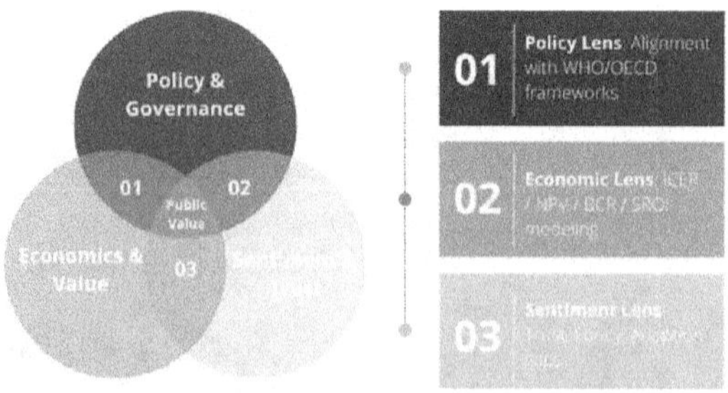

Fig 2. The "Tri-Lens Assessment Framework"

1.1 Purpose and Approach

We examine how ABDM's foundational DPI (ABHA, HPR, HFR, PHR, NHCX) enables integration, efficiency, equity, and trust. Methods included:

- Policy & comparative analysis against WHO's pillars and OECD readiness dimensions.
- Economic modelling to link qualitative frameworks with ICER/NPV/BCR/SROI and a supply-side affordability threshold (λ_IN ≈ ₹14,000–₹22,000 per QALY), derived from public healthcare spend per capita (Indian Reference Case/HTAIn; departmental budget data).
- A five-pillar evaluation (Strategy & Governance; Infrastructure & Interoperability; Equity & Access; Service Delivery & Innovation; Monitoring & Impact).
- Directed content/sentiment analysis of stakeholder interviews using six codes (Trust_Infra, Friction_Adoption, Risk_Privacy, Hope_Transform, Apathy_Disengaged, Equity_Gap).

1.2 What the Evidence Shows

1.2.1. Strategic design is globally aligned BUT co-ordination is the weak link

ABDM's vision, governance locus (NHA), standards (HL7 FHIR), and consent architecture are consistent with the WHO/OECD doctrine and peer exemplars (UK, Singapore, Estonia). However, inter-ministerial and centre-state co-ordination is uneven and relies heavily on administrative will rather than codified mechanisms (NHP 2017; NDHB 2019; WHO 2020–25; OECD 2023).

1.2.2. Robust digital infrastructure BUT limited interoperability at point of care

ABDM's technical foundations are among the most advanced globally in scale and architecture. By mid-2025, over 620 million ABHA IDs, 200,000 registered facilities, and 250 million linked health records had been created. Core registries are functional, and the open API ecosystem is active. Yet, real-world interoperability- especially at the primary and secondary care levels- remains constrained. Only a minority of facilities consistently generate and exchange FHIR- compliant data. Provider-side digital capacity, integration costs, and limited incentives for ABDM adoption at the point of care are recurring barriers.
International comparators (e.g., Singapore, the UK, and Estonia) demonstrate that technical standards must be coupled with mandatory compliance and usage-linked incentives to translate digital infrastructure into service-level efficiency.

1.2.3. Progress on access BUT equity and literacy gaps persist

ABDM's expansion has been largely urban-driven, reflecting underlying disparities in digital literacy, connectivity and device access. Rural and low-income populations, particularly those served by primary health centers, remain underrepresented in digital participation. While over 600 million citizens have an ABHA ID, active utilization and PHR linkages are significantly lower in underserved regions. Without assisted digital mechanisms, the risk of digital exclusion may amplify existing inequities. Bridging this gap will require a dedicated focus on digital literacy, language localization, and human-assisted enrollment models, aligned with India's broader inclusion strategies under the Digital India programme.

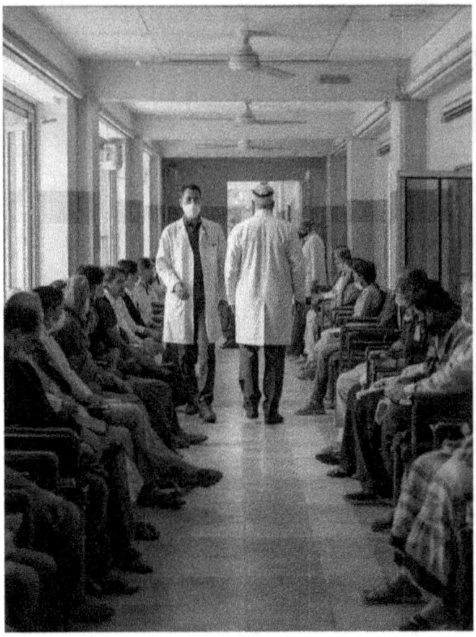

1.2.4. Early signs of value creation BUT outcomes linkage is nascent

The economic potential of ABDM is evident but under-realized. The digitization of registries and claims processes has demonstrated reductions in administrative burden and potential for improved service efficiency. Pilot analyses in selected states indicate time savings in claims settlement and improved traceability of clinical encounters. However, measurable linkage between ABDM utilization and clinical or financial outcomes- such as reduction in duplicate diagnostics, improved care continuity, or decreased out-of-pocket expenditure- remains limited.

International experience (OECD, WHO, World Bank) suggests that the transition from infrastructure to outcomes requires robust monitoring frameworks with outcome-oriented key performance indicators (KPIs). India's current dashboards are largely input-focused (e.g., number of ABHA IDs, facilities onboarded). A next-generation monitoring system that includes care-quality, safety, and financial-protection indicators would strengthen the economic and social case for sustained investment.

1.2.5. Cybersecurity and trust are critical enablers BUT need enforcement

Legal and policy frameworks for data protection are advancing, anchored by the Digital Personal Data Protection Act (DPDP) 2023. However, cybersecurity readiness across health facilities- particularly smaller institutions and startups- lags behind. Establishing a sectoral health CERT, implementing minimum cybersecurity controls for ABDM-linked systems, and conducting regular audit and breach simulations would align India with international good practice and strengthen stakeholder trust.

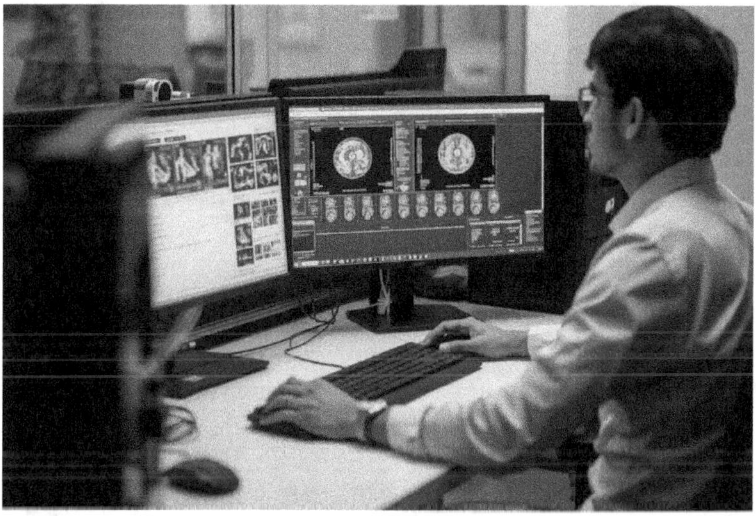

1.3 Economic and Policy Implications

ABDM represents a long-term investment in digital public infrastructure with substantial potential fiscal and societal returns.

- Utilisation—not enrolment—drives returns. The fiscal and social ROI from fewer duplicate tests, faster claims and better continuity accrues only with regular use at point of care.
- Adoptability is a policy instrument. International experience shows that standards need compliance levers (accreditation/empanelment/reimbursement) and user-facing benefits (time saved, less paperwork) to sustain adoption that will generate positive benefit-cost ratios (BCR) and social return on investment (SROI) (OECD 2023; WHO 2020–25).
- Affordability thresholds must be India-specific. Use λ_IN (updated QALY Baseline presented in this paper) for procurement, pilots, and scale decisions; keep GDP-linked thresholds as advocacy/upper-bound sensitivity only (HTAIn; Indian budget documents).

1.4 Priority Actions 2025 - 2027

INSTITUTIONALISE CO-ORDINATION

Set up an Inter-Ministerial Digital Health Council (NHA–MoHFW–MeitY–States) with statutory reporting lines; publish a single cross-ministerial performance dashboard. Also, harmonize legacy state digital health systems with ABDM architecture through funding and technical assistance for states.

01.

SHIFT KPIS FROM INPUTS TO OUTCOMES

Add care- quality/safety, continuity of care, financial-protection (OOPE/catastrophic spend), and PROMs to ABDM's core metrics; link programme reviews to these indicators (WHO M&E). Introduce incentive schemes for providers demonstrating verified improvements in efficiency, claims turnaround, or reduced duplication.

02.

MANDATE INTEROPERABILITY AT THE POINT OF CARE

Enforce FHIR-compliant referrals, discharges, e-prescriptions for ABDM-connected transactions; tie to NABH/NQAS accreditation and PM-JAY empanelment/claims.

03.

LAUNCH A DIGITAL HEALTH INCLUSION MISSION

Finance 'assisted-use' models (Tech-ASHA), multilingual/offline-first PHR, device/connectivity support for low-resource facilities, and public literacy campaigns; stratify dashboards by gender, geography, and income.

04.

EMBED ECONOMICS AND PUBLISH VALUE

Stand up an ABDM Economic Dashboard (ICER, NPV, BCR, SROI) using λ_IN; run 2-3 rapid HTA-linked pilots (e.g., NCD e-referrals, claims automation) and publish value-for-money results.

05.

OPERATIONALISE CYBERSECURITY AND TRUST

Create a Health-CERT under NHA; mandate minimum security controls, annual public cyber-audits, and breach-response statistics; publish a Digital Health Trust Index for ABDM-connected solutions.

06.

Fig 3. Priority Actions for Strengthening ABDM Implementation

1.5 Conclusion

ABDM has moved India from policy maturity to architectural maturity. The next phase must deliver performance maturity- measurable gains in access, quality, and financial protection—equitably and securely. If governance, incentives, economics, and cyber-trust are institutionalised alongside technology, ABDM can evolve into a national learning health system and a global benchmark for federated, inclusive digital health (NHP 2017; NDHB 2019; WHO 2020–25; OECD 2023; DPDP 2023; NHA 2022–23; ABDM Dashboard, 19 Aug 2025).

2. Background & Problem Statement

2.1 Background

India's health system is one of the world's largest and most mixed, combining a vast public network with a highly heterogeneous private sector that delivers a large share of outpatient and inpatient care. Policy reform over the past decade has focused on two parallel goals: advancing universal health coverage (UHC) and modernising health information infrastructure. The National Health Policy (NHP) 2017 explicitly positioned digital health as a core enabler of UHC- calling for interoperable systems, privacy-by-design, and a federated architecture suitable for India's federal structure and mixed delivery model (NHP 2017). The National Digital Health Blueprint (NDHB) 2019 translated that vision into a technical and governance blueprint- defining building blocks, registries, standards (HL7 FHIR), consent artefacts, and institutional roles (NDHB 2019).

The Ayushman Bharat Digital Mission (ABDM), launched in 2021 under the National Health Authority (NHA), operationalises this blueprint. ABDM's core components- ABHA (unique health IDs), Health Facility Registry (HFR), Health Professional Registry (HPR), Personal Health Records (PHR), and the National Health Claims Exchange (NHCX)- are designed to enable secure, consented, standards-based exchange across public and private providers, payers, and citizens (NHA Annual Report 2022–23; ABDM Building Blocks; NRCeS FHIR Implementation Guide 2024). India's broader digital public infrastructure (DPI) experience (e.g., Aadhaar, UPI, DigiLocker) informed ABDM's design choices: open standards, public-private ecosystem participation, and scalable APIs.

This digital programme sits alongside major coverage and service reforms- most notably Ayushman Bharat PM-JAY (publicly financed hospital insurance for low-income families) and e-health services such as eSanjeevani (telemedicine). Fiscal and utilisation trends provide important context: government health expenditure has increased as a share of total health spending while out-of-pocket expenditure has fallen over the last decade; yet affordability and access gaps persist for large segments of the population, especially in primary care and rural settings (budget documents; National Health Accounts; programme reports). For international readers and private-sector stakeholders, ABDM should be understood as national digital infrastructure- not a single app or scheme- intended to make health data portable, programmable, and privacy-respecting across a diverse delivery landscape.

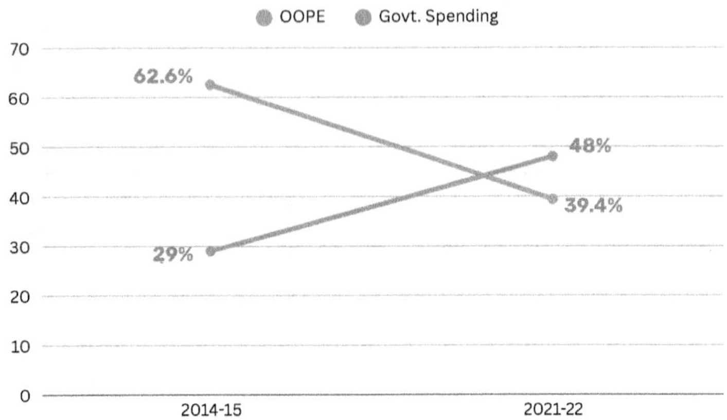

Fig 4: Govt. spending in healthcare has crossed out-of-pocket expenditure (OOPE) in 2021-22

2.2 Problem Statement

Despite rapid architectural progress, India now faces a second-order challenge: converting digital infrastructure into equitable, measurable public value. This matters now because India has reached architectural maturity; the constraint is no longer whether digital systems can be built, but whether they can reliably deliver value- at the bedside, at the front desk, and at claims desks- for everyone. Moving from policy readiness to performance readiness requires: codified co-ordination, adoption incentives linked to core transactions, equity-by-design implementation, outcome-oriented KPIs, routine economic tracking using λ_IN, and sector-grade cyber assurance. The remainder of this whitepaper examines these issues through policy benchmarking, economic methodology and results, and stakeholder sentiment, and then proposes a pragmatic roadmap to institutionalise value creation at the national scale.

Four interlinked problems define this transition:

2.2.1 Co-ordination and Accountability in a Federated System

Institutional roles are clear on paper- NHA as platform steward; states as implementers; providers and technology partners as ecosystem participants- but horizontal co-ordination (across ministries and programmes) and vertical integration (centre-state-facility) are uneven. Legacy state systems run in parallel with varying alignment to ABDM standards; the enforcement of interoperability and adoption responsibilities is still maturing. The result is variable uptake and a diffusion of accountability for outcomes beyond enrolment (NHP 2017; NDHB 2019; NHA reports).

2.2.2 Adoption Depth at the Point of Care

Core registries and APIs are live and scaling, yet routine, FHIR-compliant data exchange in everyday workflows- referrals, discharges, e-prescriptions, diagnostics- remains inconsistent, especially in primary/secondary care and smaller facilities. Providers cite integration costs, UI/UX and workflow fit, training needs, and limited "what's-in-it-for-me" benefits. International data indicates that technical standards achieve impact only when paired with usage-linked incentives (e.g., accreditation, empanelment, reimbursement) and visible time/efficiency gains (WHO 2020–25; OECD 2022/23; NHA Annual Report 2022–23).

2.2.3 Equity, Literacy, and Inclusion Risks

Digital enrolment has outpaced active, meaningful use in underserved geographies. Digital literacy, language/localisation, connectivity, and assisted-use are binding constraints for citizens and frontline staff alike. Without targeted inclusion measures- multilingual/offline-first tools, community-based facilitation (e.g., Tech-ASHA models), and low-friction provider onboarding- ABDM risks replicating or amplifying existing inequities across gender, income, and rural–urban lines (ABDM dashboard trends; Digital India inclusion materials).

2.2.4 Measurement, Economics, and Trust

Current dashboards emphasise inputs (IDs created, facilities onboarded) more than outcomes (continuity of care, safety, financial protection). Economic analyses indicate that digital interventions can be cost-effective within India-specific affordability thresholds (λ_IN, derived from public health expenditure per capita), and that benefits such as reduced duplication and faster claims can yield positive NPV/BCR/SROI. Yet these are not institutionalised in routine monitoring or budgeting cycles. In parallel, while the DPDP Act 2023 provides a legal scaffold for data protection, operational cybersecurity (sectoral CERT capacity, minimum controls, audits, drills) and transparent data-use communication are now essential to sustain provider and citizen trust (HTAIn/Indian Reference Case; DPDP 2023; programme security guidance).

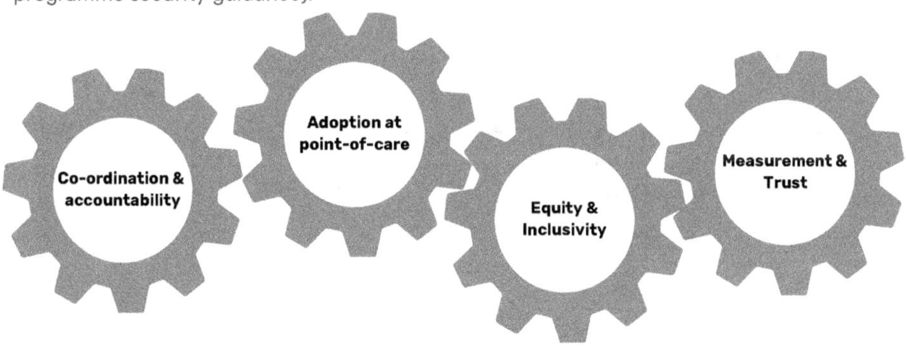

Fig 5: India faces four key linked problems in implementation of a universal digital health programme

3. Policy & Global Benchmark Review

3.1 Purpose and Scope

This section evaluates whether India's Ayushman Bharat Digital Mission (ABDM) aligns with internationally recognised principles of digital-health governance, interoperability, and value-realisation. The analysis benchmarks India's policy trajectory and implementation maturity against the WHO Global Strategy on Digital Health 2020-2025, the OECD Digital Health Readiness Framework, and national exemplars including the UK (NHS Digital/NHS England), Singapore (NEHR), Estonia (X-Road), and Australia (My Health Record).

Sources include:

- Government of India policy artefacts (NHP 2017, NDHB 2019, ABDM Notifications 2021, DPDP Act 2023)
- NHA operational documents and dashboards (2022-25)
- Standards and guidance (NRCeS FHIR Implementation Guide 2024, QCI/NABH vendor certification, DHIS guidelines)
- Comparative frameworks and country reports from WHO, OECD, and World Bank.

The review builds and applies a five-pillar evaluation framework developed based on the WHO guidance for digital health.

1. Strategy & Governance
2. Infrastructure & Interoperability
3. Equity & Access
4. Service Delivery & Innovation
5. Monitoring & Impact

Each source was mapped against these pillars and scored (1 = very limited → 5 = mature) to derive comparative insights.

3.2 Evolution of India's Digital-Health Policy

India's digital-health ecosystem has evolved through three deliberate policy phases. The result is a technically advanced yet federated system that balances innovation and sovereignty- distinct from the centralised architectures of smaller nations.

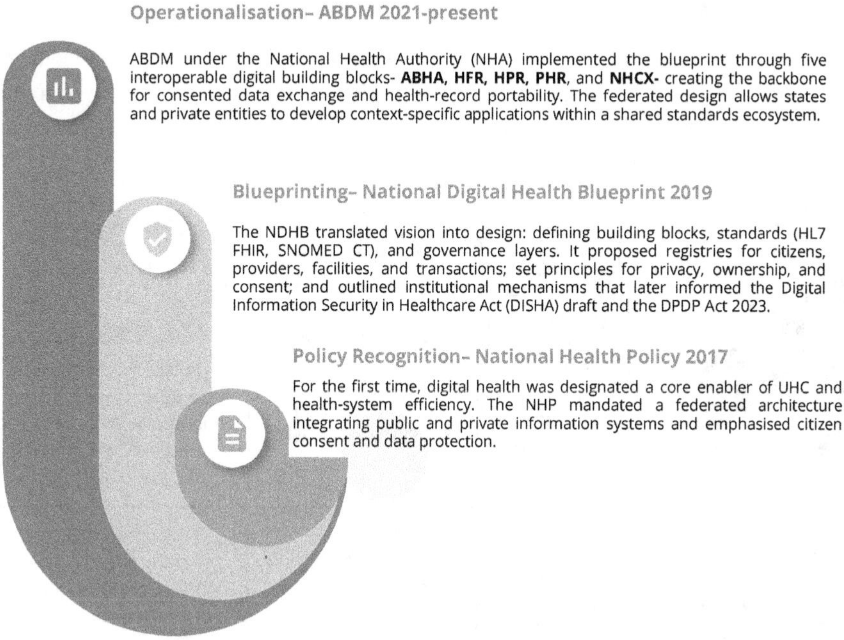

Fig. 6: Evolution of India's Digital Health Ecosystem

3.3 Global Frameworks & Case Comparators

WHO Global Strategy 2020–2025

Defines four strategic objectives:

1. Strengthen governance and leadership
2. Promote national digital health strategies and capacity
3. Advance interoperability and data standards
4. Enable person-centred care and data trust

India aligns strongly with objectives 1 and 2, and while it has policies around 3, enforcement is lacking. This has been done through NHA stewardship and open-standards mandates, but is still transitioning from capacity building to measurable outcomes associated with objective 4.

OECD Digital-Health Readiness Indicators (2019-22)

Benchmarks infrastructure maturity, governance, adoption, and impact monitoring. India scores high on infrastructure and standards for a middle-income economy but moderate on data quality, secondary use, and evaluation capacity. They have a higher level of quantitative and qualitative frameworks to track nations and regions.

Comparator	Distinctive Feature	Key Learning for India	What Cannot Be Directly Applied (and Why)
UK (NHS Digital)	Centralised repositories, mandatory EHR standards linked to funding	Enforcement and alignment of finance with interoperability targets	Full centralisation of records is infeasible due to India's federal diversity and fragmented provider landscape
Singapore (NEHR)	Gradual roll out with strong patient consent controls	Incremental implementation and public trust campaigns	Singapore's scale and uniform infrastructure cannot be replicated; India needs multi speed adoption for rural and urban contexts
Estonia (X Road)	Legal backbone and cybersecurity institutionalised early	Integrate legal and technical assurance from the outset	Estonia's population size and digital penetration enable simplicity; India must design for massive heterogeneity and low bandwidth environments
Australia (My Health Record)	Opt out architecture with high engagement	Importance of communication and opt out design for scale	Opt out model may not suit India due to literacy variation, cultural diversity, and the need for stronger consent safeguards

India's model is unique in its scale and federal diversity- combining mandatory standards with voluntary ecosystem participation. This approach offers flexibility but requires stronger mechanisms for compliance and performance incentives.

3.4 Findings from Comparative Review

> India's digital health strategy shows rare continuity and ambition, anchored in a federated architecture built for scale. Infrastructure has grown fast, but adoption depth and outcome linkage remain uneven. Privacy principles are largely aligned with global norms, while cyber-assurance capacity must catch up. The next inflection depends on shifting from process metrics to measurable care, cost and equity outcomes.

3.4.1. Strategic continuity with governance complexity.

Policy coherence across NHP → NDHB → ABDM demonstrates institutional learning and vision stability. However, execution depends on inter-ministerial co-ordination (MoHFW, MeitY, Finance, Insurance) and state-level capacity. Decisions remain personality-dependent rather than codified through law or regulation..

3.4.2. Federated architecture is India's differentiator.

Unlike centralised models, the federated design permits context-specific innovation while maintaining core standards. The trade-off is complex accountability: without uniform standards enforcement and financial alignment, interoperability can remain technical rather than functional. This is needed in a country as large, diverse and populated as India.

3.4.3. Rapid infrastructure scale, moderate adoption depth.

Registry growth and API use are strong, but real-world use for care continuity and decision support is still limited. Global comparators achieved impact when adoption was linked to provider payments and licensing.

3.4.4. Privacy alignment, cyber preparedness lag.

The DPDP Act 2023 aligns India with OECD principles on consent and purpose limitation. Operational cybersecurity (CERT capacity, audits, breach response) is developing. Estonia and Australia show that robust cyber assurance underpins trust and data use.

3.4.5. Outcome monitoring remains process-centric.

ABDM tracks enrollment and integration metrics but not yet care-quality, efficiency, or equity outcomes. OECD and WHO frameworks define such indicators as essential for evidence-based funding and public accountability.

3.5 Comparative Scoring Summary

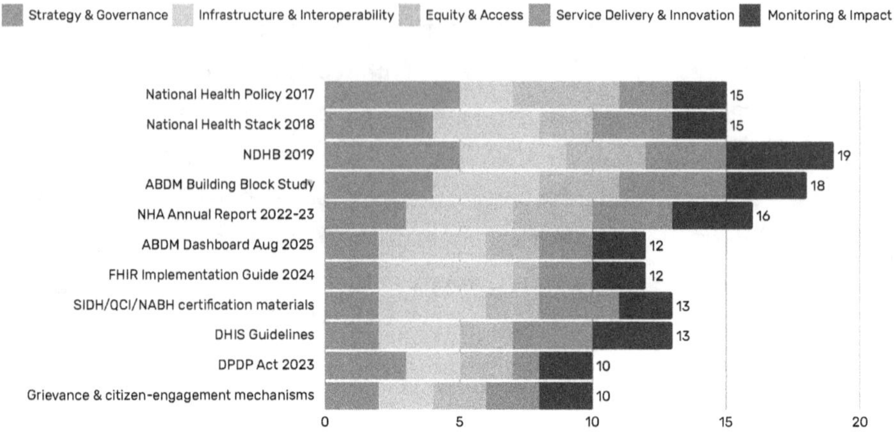

Fig. 7: Scoring summary

Pillar	Average Score (out of 5)
Strategy & Governance	3.0
Infrastructure & Interoperability	3.6
Equity & Access	2.3
Service Delivery & Innovation	2.6
Monitoring & Impact	2.4

**The scores in this table represent average ratings across all sub indicators within each pillar, using a 1 to 5 scale derived from the ABDM readiness assessment framework. Each pillar (Strategy and Governance, Infrastructure and Interoperability, Equity and Access, Service Delivery and Innovation, and Monitoring and Impact) contains multiple qualitative and quantitative elements that were independently rated based on evidence from policy documents, operational data, and stakeholder inputs. The values shown here are the mean scores, which provide a simplified view of relative maturity levels across pillars.*

> When tested, the framework showed that ***India's infrastructure and standards maturity rank among the global leaders***, while **equity integration, routine service adoption, and outcome measurement continue to lag behind-** a pattern consistent with the qualitative findings and comparative reports.

3.6 Synthesis and Implications

3.6.1. Structural readiness > adoption maturity.

India has built a solid scaffold- registries, APIs, and governance protocols- comparable to high-income peers. The critical next step is to embed these within care delivery and financing workflows.

3.6.2. Interoperability must become behavioural

Providers respond to aligned incentives and feedback loops. Mandatory standards and value-linked rewards are key to turning technical capacity into everyday use.

3.6.3. Trust and inclusion drive sustainability

Legal protections and federated design create flexibility but require visible accountability and cyber readiness to retain stakeholder confidence.

3.6.4. Monitoring must capture impact, not inputs.

Outcome-linked KPIs- safety, continuity, financial protection- should be institutionalised in NHA dashboards and budget reviews.

Together, these findings place ABDM at a pivot from policy maturity to performance maturity. Its next milestone is institutionalising economic and social value at the same scale as its digital infrastructure.

4. Economic Methodology & Results

4.1 Rationale and Objectives

Digital-health transformation requires not only technical and policy validation but also economic justification- evidence that investments in platforms like ABDM generate measurable health and fiscal returns. Historically, Indian evaluations have relied on input metrics (numbers of IDs or facilities onboarded), whereas long-term value-for-money analysis- standard in OECD systems- has been limited. This section establishes a structured **Economic Prioritisation Model (EPM)** that adapts global health-economics principles to India's fiscal reality, allowing both policymakers and investors to assess whether ABDM interventions are affordable, cost-effective, and socially beneficial.

4.2 Framework and Models

4.2.1 Core Analytical Principles

Four complementary models- each answering a different policy question- were applied:

MODEL	DEFINITION	POLICY USE
ICER - Incremental Cost-Effectiveness Ratio	Measures the additional cost per incremental health benefit (e.g., ₹/QALY gained).	Determines whether an intervention is "worth it" relative to national willingness-to-pay.
NPV - Net Present Value	Calculates discounted net fiscal benefit over time: future savings- current cost.	Tests long-term budget sustainability.
BCR - Benefit-Cost Ratio	Ratio of total (discounted) benefits to total costs.	Communicates efficiency in simple terms for budget decisions.
SROI - Social Return on Investment	Monetises non-financial outcomes such as equity, access, and time saved.	Integrates social value and inclusion into investment logic.

These were mapped against the five evaluation pillars (Strategy & Governance, Infrastructure & Interoperability, Equity & Access, Service Delivery & Innovation, Monitoring & Impact) to ensure multidimensional assessment rather than purely fiscal scoring.

4.3 Setting an India-Specific Threshold (λ_IN)

4.3.1 Why GDP-Based Thresholds Fail

The HTAIn Reference Case (2018) endorses QALY as the preferred outcome unit and cites a GDP-based threshold of ≈ ₹2.1 lakh/QALY (1× GDP per capita). However, GDP-linked thresholds overstate real affordability for a publicly financed system where government expenditure constitutes only ~1.3 % of GDP. They ignore fixed budget envelopes and the marginal productivity of existing health spend.

4.3.2 Derivation of λ_IN

To estimate a practical, supply-side threshold:

1. Total public health expenditure ≈ ₹2.9 trillion (Union + State).
2. Population ≈ 1.4 billion → ₹ 2,070 per capita per year.
3. Assuming 1 QALY corresponds to one healthy life-year and reallocating ~0.7 % of spend to cost-effective innovations, the implied affordability band becomes ₹ 14,000–₹ 22,000 per QALY.

This λ_IN therefore represents India's budget-feasible marginal productivity threshold, distinct from aspirational GDP metrics. It serves as the base reference for all model interpretations, with the GDP value retained only for sensitivity testing.

Illustrative Example to calculate QALY

Assume person X suffers from a health condition, experiencing different health states over the years

Time Period	Health	Utility Weight	Years	QALYs
Year 1	Good health with mild symptoms	0.8	1	0.80
Years 2–3	Moderate chronic symptoms	0.6	2	1.20
Year 4	Near-full recovery	0.9	1	0.90

$$\text{QALY} = \Sigma\,(\text{Utility} \times \text{Time}) = (0.8 \times 1) + (0.6 \times 2) + (0.9 \times 1)$$

Total QALY for Person X = 2.9

4.4 Illustrative Calculations

Model	Formula	Illustrative ABDM Use-Case	Result & Interpretation
ICER	ΔCost / ΔQALY	Avoiding one duplicate hospitalisation (₹ 25000) yields 0.15 QALYs.	ICER = ₹ 1.67 lakh/QALY → Cost-effective vs GDP threshold; not vs λ_IN → shows the importance of India-specific λ.
NPV	Σ (Benefits - Costs)/(1 + r)^t	₹ 500 cr investment → ₹ 100 cr annual savings for 10 yrs, r = 5 %.	PV benefits = ₹ 772 cr → NPV +₹ 272 cr, fiscally sustainable.
BCR	PV Benefits / PV Costs	Same scenario.	BCR = 1.8 → Every ₹ 1 invested returns ₹ 1.80.
SROI	(Social Value - Investment)/Investment	₹ 1500 cr social value from ₹ 500 cr investment (equity, time saved).	SROI = 2.0, a 200% social return.

Together, these illustrate that ABDM-type investments can be both economically and socially sound, provided utilisation remains high and equity externalities are captured.

4.5 Application within the Five-Pillar Framework

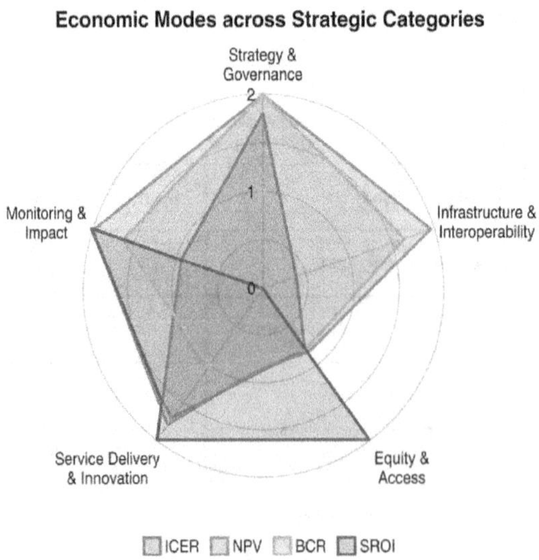

Fig. 8: Star graph representation of economic modes across strategic categories

Economic Mode	Strategy & Governance	Infrastructure & Interoperability	Equity & Access	Service Delivery & Innovation	Monitoring & Impact
ICER	Informs prioritisation of strategic interventions	--	--	Measures the efficiency of clinical/digital interventions	Tracks outcome gains (QALY/DALY)
NPV	Justifies long-term fiscal sustainability	Captures savings from interoperable infra	--	Guides innovation investment	Compares projected vs realised benefits
BCR	Communicates governance-level efficiency	Summarises the ROI of infra projects	--	Applies to pilots with clear cost/benefit	Enables simple reporting of efficiency
SROI	--	--	Captures access, equity, and inclusion	Highlights social innovation value	Tracks OOPE reduction and trust metrics

4.6 Empirical Insights

4.6.1. Threshold Alignment

Many ABDM components- digital claims, registry integrations, e-referrals- are cost-effective even under conservative λ_IN assumptions.

4.6.2. Time Horizon Sensitivity

High upfront costs yield positive NPV beyond 3–5 years, arguing for sustained fiscal commitment

4.6.3. Equity Multiplier

When equity and trust are monetised, SROI doubles total return; inclusion thus strengthens- not dilutes- economic justification.

4.6.4. Fiscal Integration

Embedding HTAIn cost-utility analysis into ABDM budgeting cycles will normalise evidence-based funding.

4.7 Policy Recommendations for Economic Institutionalisation

- **Adopt λ_IN (₹ 14k–₹ 22k/QALY)** as a reference for national and state-level health-tech procurement.

- **Create an ABDM Economic Dashboard** combining ICER, NPV, BCR, SROI to monitor programme efficiency.
- **Integrate HTAIn** review into all major ABDM projects before scale-up.
- **Publish annual "Digital Health Value Reports"** to communicate efficiency and social returns transparently.
- **Train state analysts in cost-effectiveness** and value-for-money modelling to enable decentralised evaluation.

4.8 Key Takeaway

*India has demonstrated that **digital health is not only technically viable but economically rational**. When evaluated against λ_IN, ABDM investments remain fiscally affordable, socially beneficial, and politically defensible. The policy task ahead is to **institutionalise these economic tools-** embedding cost-utility thinking into every layer of planning, budgeting, and accountability.*

4.9 Summary of Economic Findings by Pillar and Metric

Aggregate Summary (Weighted Across Pillars)

- Mean BCR = 1.84
- Mean SROI = 2.03
- Composite Economic Readiness Index = 0.72 (out of 1) → strong structural efficiency, moderate adoption elasticity.

Interpretation:

- ABDM demonstrates economic viability even under conservative affordability assumptions.
- Equity and monitoring investments, often seen as soft expenditures, generate the highest combined SROI + BCR gains.
- Infrastructure returns are delayed but stable- underscoring the need for long-term fiscal commitment rather than short project windows.

Pillar	Strategy & Governance	Infrastructure & Interoperability	Equity & Access	Service Delivery & Innovation	Monitoring & Impact
Economic Model(s) Applied	NPV + BCR	ICER + NPV	SROI + BCR	ICER + SROI	NPV + SROI
Primary Evidence / Illustrative Use-Case	Governance reforms and digital-platform rollout (NHA + State systems)	Health-facility digitisation and FHIR-based registry expansion	Tech-ASHA models, PHR multilingual interface, rural tele-health pilots	Digital claims, AI-screening, e-referral modules	Outcome dashboards and evaluation capacity building
Result (vs λ_IN ₹14–22 k/QALY)	NPV +₹ 272 cr; BCR = 1.8	₹ 1.6 lakh/QALY (GDP-cost-effective; ≈ 7× λ_IN)	SROI = 2.3 ; BCR = 1.5 > 1	ICER ₹ 10.8 k/QALY < λ_IN → cost-effective	BCR = 1.6; SROI = 1.8
Fiscal / Social Implication	Medium-term fiscal savings through reduced duplication in IT systems	Large upfront capex → positive NPV after 4 years of utilisation	Every ₹1 → ₹ 2.3 in social value (time saved, access gains)	Operational savings ≈ 25 % per claim cycle	Moderate fiscal gain + significant accountability value
Policy Message	Co-ordination and central dashboards deliver tangible fiscal return.	Infrastructure viable if sustained funding bridges initial negative ROI phase.	Equity investments are fiscally positive when measured in social ROI, not pure ICER.	Innovation spend is efficient under India-specific thresholds.	Institutional evaluation is a positive investment, not an overhead.

5. Stakeholder Sentiment Analysis

5.1. Objective and Context

Understanding stakeholder sentiment is essential for assessing functional adoption and institutional legitimacy- two determinants not captured by technical or economic metrics. While policy and economic analyses measure capacity and value, sentiment analysis reveals the confidence gap between design and lived experience. This section summarises findings from a structured qualitative review- N= 20 semi-structured interviews- conducted between March 2024 and July 2025 across public, private, and civil-society stakeholders.

Participants represented:

- 6 public-sector health administrators (national and state level)
- 5 private hospitals and health-tech firms
- 4 frontline providers (PHCs, district hospitals)
- 5 insurers/TPAs and 2 civic-technology or patient-rights organisations

5.2 Methodology

5.2.1 Coding Framework

A directed content-analysis model was applied using six pre-defined codes derived from the whitepaper's conceptual framework and the WHO/OECD digital-readiness constructs

Each interview transcript was manually coded by two reviewers. Divergent codes were reconciled through consensus, producing an overall Net Sentiment Index (NSI) for each dimension.

5.2.2 Validation

Findings were triangulated with published grey literature- public consultations, ABDM webinars, industry whitepapers (NASSCOM 2024), and press statements- to ensure consistency and exclude anecdotal bias.

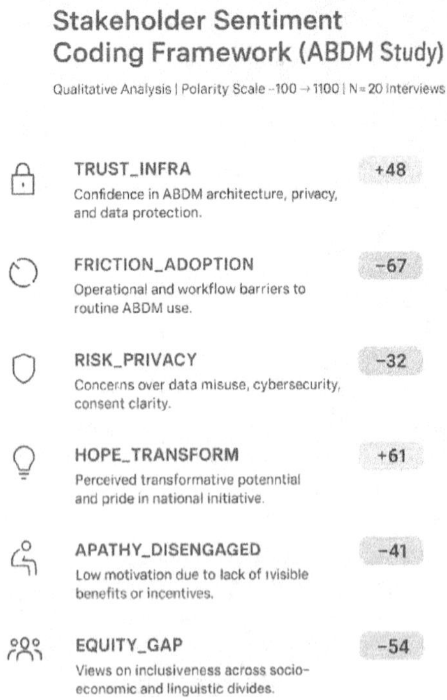

Fig. 9: Stakeholder Sentiment Analysis

5.3 Findings

5.3.1. Trust_Infra- Moderate-Positive (+48)

Respondents expressed confidence in the NHA's institutional credibility, referencing successful platforms like CoWIN and eSanjeevani as precedents. The Digital Personal Data Protection (DPDP) Act 2023 was cited as a trust anchor, though operational transparency on breach response and grievance handling remains limited.
Private providers appreciated the open-API approach and voluntary sandbox participation but wanted clearer certification pathways for ABDM-linked software vendors.

5.3.2. Friction_Adoption- Strong-Negative (-67)

Operational friction emerged as the single largest deterrent. Providers cited:

- Redundant data entry across multiple systems
- Slow internet or unstable network connections in rural PHCs
- Unclear workflows for FHIR-based referrals and discharges
- Administrators noted the absence of routine incentives or time-saving benefits: "We comply because it's mandatory, not because it helps the patient encounter."

5.3.3. Risk_Privacy - Moderate-Negative (-32)

While high-level legal assurances exist, on-ground comprehension of consent mechanisms is weak. Only one-third of facilities interviewed had formal privacy-training modules. Fear of reputational damage in the event of breaches discourages open data-sharing even within authorised frameworks

5.3.4. Hope_Transform - Strong-Positive (+61)

Across stakeholder categories, there was strong optimism that ABDM could "do for health what UPI did for finance." Respondents highlighted national pride, scale, and interoperability potential. Technology firms viewed ABDM as a "public digital platform with private innovation opportunities."

5.3.5. Apathy_Disengaged - Moderate-Negative (-41)

Among mid-level facility managers and public-sector clinicians, digital fatigue was common. Previous uncoordinated e-initiatives (HMIS, RCH, NIKSHAY) had created scepticism. Many perceive ABDM as "another portal," unless accompanied by workflow simplification or direct patient benefits.

5.3.56. Equity_Gap - Significant-Negative (-54)

Respondents consistently cited urban-rural and income-based digital divides. Only a minority of facilities reported multilingual interfaces or assisted-use kiosks. Private sector executives acknowledged the business case for inclusivity but prioritised profitable, connected markets first. Civil-society actors stressed that without assisted models (Tech-ASHA, offline PHR), digital health could exacerbate exclusion.

5.4 Cross-Code Correlations

Trust_Infra ↔ Hope_Transform (+0.78): Confidence in governance directly strengthens optimism for transformation.

Friction_Adoption ↔ Apathy_Disengaged (+0.83): Operational inefficiency drives disengagement.

Equity_Gap ↔ Risk_Privacy (+0.56): Populations least digitally literate are also least aware of consent and rights.

Trust_Infra ↔ Risk_Privacy (−0.59): Breach anxieties erode system trust despite legal assurances

This matrix underscores that trust and usability are reinforcing, whereas friction and exclusion form a mutually amplifying negative loop.

5.5 Sentiment Heat Map Summary

Dimension	Mean Sentiment Score	Trend vs 2024 Baseline	Directional Insight
Trust_Infra	+48	↑ +9	Confidence rising post-DPDP 2023 and NHA transparency drives.
Friction_Adoption	-67	↓ -7	Persistent workflow friction; little improvement in PHC settings.
Risk_Privacy	-32	↑ +5	Awareness growing, but clarity gaps remain.
Hope_Transform	+61	↑ +4	Optimism sustained; branding success of ABDM visible.
Apathy_Disengaged	-41	↔ 0	Unchanged; incentive deficit unaddressed.
Equity_Gap	-54	↓ -6	Urban–rural disparity widening slightly.

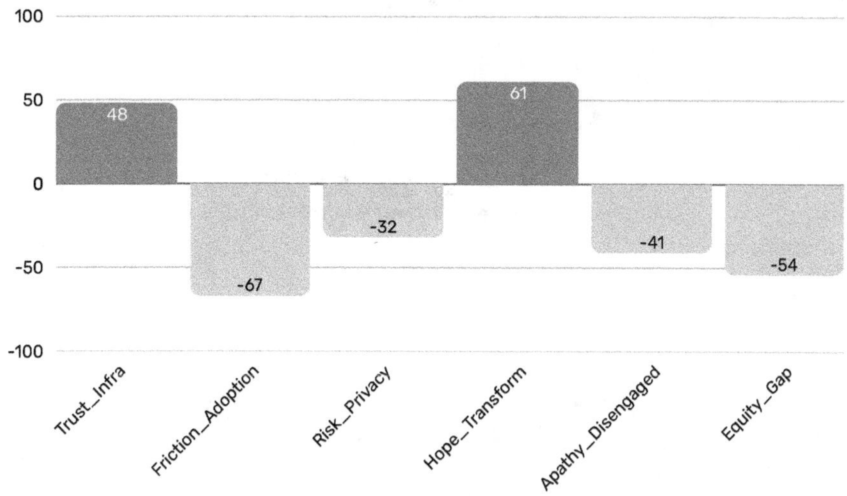

Fig 10: Summary of Stakeholder Sentiment Scores

Overall Net Sentiment Index (NSI) = -14 → Cautious optimism, skewed by operational and equity concerns.

5.6 Key Insights and Policy Implications

- **Institutional trust is emerging as ABDM's strongest intangible asset,** but it is fragile; transparency in data use and breach response will determine whether trust translates into sustained engagement.

- **Operational friction is the most immediate barrier**- digital workflows must deliver time savings and visible user value to reverse negative sentiment.

- **Equity remains the decisive variable.** Stakeholder empathy and user assistance (Tech-ASHA, local-language PHR) can convert negative sentiment into adoption momentum.

- **Sentiment data could be institutionalised**- embedding structured perception tracking into NHA dashboards alongside fiscal and clinical KPIs to guide adaptive governance.

6. Integrated Insights & Strategic Recommendations

6.1 Synthesis of Evidence

Across policy benchmarking, economic evaluation, and stakeholder sentiment analysis, a consistent picture emerges: **ABDM is institutionally strong but operationally uneven.** Its federated architecture, data standards, and legal scaffolding match global best practice, yet on-ground adoption, co-ordination, and equity integration lag behind.

The combined results reveal five systemic dynamics:

Dimension	Empirical Finding	Interpretation
Governance	Strategic vision clear; implementation dispersed across ministries and states.	ABDM functions as infrastructure, but lacks codified cross-ministerial accountability.
Economics	Cost-effective under λ_IN (₹14–22 k/QALY); positive NPV and SROI > 1.8.	Fiscal sustainability proven-provided utilisation, not enrolment, drives value.
Adoption & Capacity	High infrastructure coverage; low point-of-care interoperability; Friction = −67 (NSI).	Workflow alignment, training, and incentive design are critical next levers.
Equity & Trust	Equity Gap= −54; Trust= +48; SROI= 2.3 when inclusion captured.	Social returns hinge on inclusion; trust acts as multiplier of adoption.
Measurement & Learning	M&E remains input-oriented.	Without outcome KPIs (quality, safety, OOPE), digital maturity cannot be proven.

6.2 Cross-Lens Insights

6.2.1 From Coverage to Consequence

Across all analyses, enrolment metrics (ABHA IDs, facility counts) overstate impact. Economic modelling shows that only active, repeated use yields fiscal and health returns. Policy focus must shift from how many joined to *how much value each use generates.*

6.3.2. From Architecture to Accountability

Institutional legitimacy now depends less on technical sophistication and more on transparent, accountable governance. A dedicated Inter-Ministerial Digital Health Council (IMDHC), legally mandated, would transform co-ordination from ad-hoc to systemic, enabling unified planning across NHA, MoHFW, MeitY, and states.

6.2.3. From Innovation to Inclusion

Sentiment and SROI analysis show that equity is an economic variable, not a moral add-on. Assisted-use models (Tech-ASHA, offline-first PHR, multilingual UI) double social ROI while strengthening trust. Inclusion should therefore be budgeted as an investment, not a subsidy.

6.2.4. From Legal Assurance to Operational Trust

DPDP 2023 created the legal floor; operational trust demands visible enforcement. Annual public cyber-audits, breach-response dashboards, and a sectoral Health-CERT will convert compliance into confidence.

6.2.5. From Reporting to Learning

To move from digital infrastructure to learning health system, ABDM must institutionalise evaluation:

- Integrate HTAIn cost-utility models into programme budgeting.
- Publish an ABDM Economic Dashboard tracking ICER/NPV/BCR/SROI.
- Add PROMs, safety, and financial-protection indicators to NHA dashboards. Measurement should evolve from counting transactions to demonstrating outcomes.

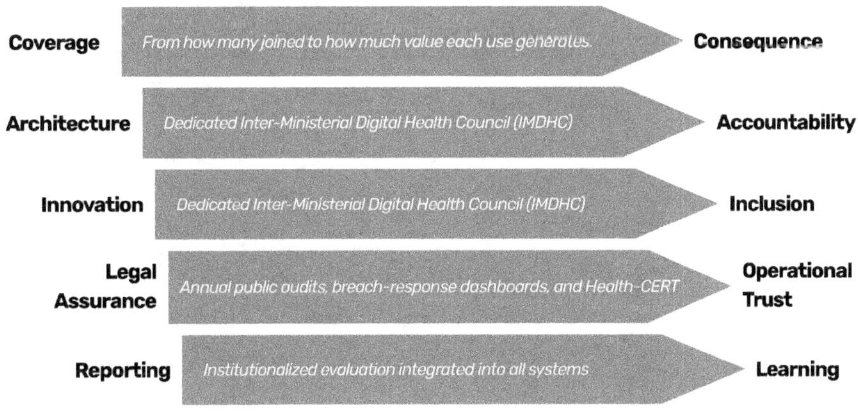

Fig 11: Transforming the Indian healthcare system

6.3 Strategic Reform Levers (2025-2030)

Lever	Objective	Illustrative Actions	Expected Outcome
Institutional Integration	Establish coherent governance	Constitute IMDHC; codify ABDM evaluation within NHA Act; align state digital-health missions	Clear lines of accountability and budget flow
Economic Institutionalisation	Embed value-for-money logic	Adopt λ_IN for cost-utility decisions; mandate HTA review for major ABDM spends	Fiscal discipline + transparent prioritisation
Equity & Access Enablement	Ensure inclusion at scale	Launch Digital Health Inclusion Mission (Tech-ASHA, rural hubs, language localization)	Higher SROI; reduced digital divide
Incentive & Behavioural Design	Shift from compliance to motivation	Tie ABDM use to NABH/NQAS, PM-JAY, and grant eligibility	Sustained adoption and data quality
Trust & Cyber Resilience	Build verifiable confidence	Create Health-CERT; annual public cyber-audit; Digital Trust Index	Trust converted into measurable assurance
Outcome-Oriented Monitoring	Institutionalise learning	Develop National Digital Health Evaluation Framework with WHO-aligned KPIs	Evidence-driven adaptation and global comparability

6.4 Reframing ABDM 2.0: The Next Transition

India's digital-health infrastructure is now complete in form but not in function. ABDM 2.0 must redefine success from "systems built" to "value delivered." Its new design philosophy should rest on five imperatives:

Fig 12: ABDM 2.0 Design Philosophy

6.5 Implementation Horizon

0-2 Y **(2025 - 2027)**	**Integration and quick wins** Unified dashboard; Health-CERT operational; first inclusion pilots launches
3-5 Y **(2027- 2030)**	**Institutionalisation and evaluation** λ_IN embedded in budgeting; Digital Health Inclusion Mission mainstreamed; outcome KPIs published
5Y+ **Beyond 2030**	**Consolidation and global benchmarking** National Digital Health Observatory; ABDM referenced as model for federated, equitable systems

7. Conclusion and Way Forward

India's Ayushman Bharat Digital Mission (ABDM) has built one of the world's most extensive public digital health infrastructures- federated, standards-based, and privacy-aware. The architecture now exists; the task ahead is to deliver tangible, trusted value for every citizen.

After four years of rapid progress, India stands at the inflection between infrastructure and impact. The technical foundations are strong and the economic case sound, but daily-use adoption, outcome tracking, and inclusion still define the distance from vision to value. The next phase must focus less on building new systems and more on making existing ones consequential:

- Every record exchanged should improve care quality or financial protection.
- Every regulation should reinforce trust through visible accountability.
- Every innovation should close- not widen- the equity gap.

In the global context, ABDM offers a new model for the Global South: federated governance with universal reach. Codifying its lessons through WHO, OECD, and South-South platforms could anchor India as the reference point for inclusive digital health transformation.

The measure of success will no longer be the number of IDs or integrations, but the number of citizens who experience safer, faster, and fairer healthcare because the system finally works- quietly, seamlessly, and for everyone.

Acknowledgements

This paper draws on conversations and exchanges across public health, clinical practice, policy, insurance, and digital health ecosystems. These interactions helped frame the Ayushman Bharat Digital Mission not only as a technological programme, but as a public system shaped by governance, incentives, and trust.

Insights from systems and policy practitioners- Dr. Rohan D'Souza, Ankur Nair, Niranjan Joshi, Jonathon Pieterse, Lt. Gen. Dr. Madhuri Kanitkar, Major Dr. Ashlesha Kelkar, Urvashi Prasad, and Kiran Anandampillai- helped anchor the analysis in questions of institutional design and long-term public value.

Frontline perspectives from Dr. Anjali Rao, Saniya Bhalerao, and Dr. Penkacherla Nagu ensured that clinical and public health realities remained central.

Analytical discussions with Dr. Andrew Street, Karanjot Chhatal, Prithwiraj Das, Arun Joshi, and Thanka Prabhu contributed to refining the policy, health economics, and digital health lens applied in this paper.

The London School of Economics (LSE), for their support through faculty insights and interviews and Viddhi Surve- a HealthEcocomics Student for her assistance during early background compilation.

Finally, Nirupa Rao for her strategic design contribution in bringing coherence to the structure and flow of the paper, helping translate complex inputs into a clear and accessible narrative.

The views expressed are solely those of the authors.

References

1. National digital health blueprint. (n.d.). Ministry of Health and Welfare, Government of India. Retrieved 14 August 2025, from https://abdm.gov.in:8081/uploads/ndhb_1_56ec695bc8.pdf
2. Ayushman bharat digital mission (Abdm)| national portal of india. (n.d.). India.Gov.In National Portal of India. Retrieved 13 August 2025, from https://www.india.gov.in/spotlight/ayushman-bharat-digital-mission-abdm
3. Palm, K., Brantnell, A., Peolsson, M., Özbek, N., & Hedström, G. (2025). National eHealth strategies: A comparative study of nine OECD health systems. BMC Health Services Research, 25(1), 269. https://doi.org/10.1186/s12913-025-12411-7
4. A brief guide on Ayushman Bharat Digital Mission (ABDM) and its various building blocks. (n.d.). National Health Authority. https://abdm.gov.in:8081/uploads/ABDM_Building_Blocks_v8_3_External_Version_eabbc5c0f3_4_a96f40c645_5716a684de_b344369144.pdf
5. Abdm-insights. (n.d.). Retrieved 16 August 2025, from https://dashboard.abdm.gov.in/abdm/
6. Global strategy on digital health 2020-2025 (1st ed). (2021). World Health Organization. https://www.who.int/docs/default-source/documents/gs4dhdaa2a9f352b0445bafbc79ca799dce4d.pdf
7. Sharma, M., Das, R., & Kapoor, A. (2024). BUDGET BRIEFS Ayushman Bharat GoI 2023-24. Accoutability Initiative, Centre for Policy Research, 15(9), 12. https://accountabilityindia.in/wp-content/uploads/2023/02/Ayushman-Bharat-Budget-Briefs-2023.pdf
8. Perappadan, B. S. (2022, February 26). Cabinet approves implementation of Ayushman Bharat Digital Mission with a budget of Rs.1,600 crore for five years. The Hindu. https://www.thehindu.com/news/national/cabinet-approves-implementation-of-ayushman-bharat-digital-mission-with-a-budget-of-rs1600-crore-for-five-years/article65086937.ece
9. NHA Annual Report 2022-23. (2023). [Annual]. National Health Authority. https://abdm.gov.in:8081/uploads/NHA_Annual_Report_2022_2023_4f3bc6e2a5.pdf
10. Implementation Guide for Adoption of FHIR in ABDM and NHCX. (2024, September). National Resource Center for EHR Standards India. https://www.nrces.in/download/files/pdf/Implementation_Guide_for_Adoption_of_FHIR_in_ABDM_and_NHCX.pdf
11. Bhandari, Y., Das, A., & Kumar, P. (2025). Digital literacy of the indian health workforce: A core requirement for healthcare delivery in the 21st century. Journal of Surgical Specialties and Rural Practice, 6(1), 9–12. https://doi.org/10.4103/jssrp.jssrp_6_25
12. Kapoor, N., Kapoor, N. T., Kumar, R., Gupta, A., Saboo, B., & Maheshwari, A. (2025). Digital foundations for health equity: Rethinking primary care through the Ayushman Bharat digital mission. Journal of Family Medicine and Primary Care, 14(7), 2605–2607. https://doi.org/10.4103/jfmpc.jfmpc_1206_25
13. Inamdar, A. V., Kale, M., Deogaonkar, A., Dariya, S. S., & Dey, A. K. (2025). A unified health ecosystem in India: Exploring the Ayushman Bharat digital mission. International Journal Of Community Medicine And Public Health, 12(8), 3813–3819. https://doi.org/10.18203/2394-6040.ijcmph20252497

References

28. *The digital personal data protection bill, 2023.* (2023, August.). PRS Legislative Research. Retrieved 5 September 2025, from https://prsindia.org/billtrack/digital-personal-data-protection-bill-2023
29. *NABH Draft Standards for HIS & EMR Standards.* (2024, July). National Accreditation Board for Hospitals ad Healthcare Providers. https://portal.nabh.co/Announcement/Draft%20NABH%20Standards%20for%20HIS%20and%20EMR%20Systems%20-%2023rd%20July%202024.pdf
30. *Nha | official website ayushman bharat digital mission.* (n.d.). Retrieved 22 August 2025, from https://abdm.gov.in/qcicertified
31. *System integrators in digital health | sidh program by national health authority(Nha).* (n.d.). Retrieved 22 August 2025, from https://www.sidh.co.in/aboutsidh'
32. Mishra, U. S., Yadav, S., & Joe, W. (2024). The ayushman bharat digital mission of india: An assessment. *Health Systems & Reform, 10*(2), 2392290. https://doi.org/10.1080/23288604.2024.2392290
33. *Digital health standards for his/emr systems—Nabh.* (n.d.). Retrieved 24 August 2025, from https://nabh.co/programmes/digital-health-standards-his-emr-systems/
34. *Digital health literacy country reports.* (2021, January.). ILA. Retrieved 24 August 2025, from https://www.ilabour.eu/results/digital-health-literacy-country-reports/
35. Banal, A. (2024, September 12). *Digital literacy and inclusion: Bridging the digital divide in india.* https://www.linkedin.com/pulse/digital-literacy-inclusion-bridging-divide-india-cyjbf
36. Basuroy, T. (2022, December 22). *India: Internet literacy index by category 2021.* Statista. https://www.statista.com/statistics/1232343/internet-literacy-index-by-category-india/
37. *Govtech maturity index (Gtmi) data dashboard.* (n.d.). [Text/HTML]. World Bank. https://doi.org/10/21/govtech-maturity-index-gtmi-data-dashboard
38. PRS Legislative Research. (2025). Demand for Grants: Health & Family Welfare – Analysis 2025-26. PRS India.
39. WHO-SEAR APO. (2022). India: Health system review (Health Systems in Transition, 11[1]). WHO.
40. WHO. (2021). Global strategy on digital health 2020–2025. WHO.
41. OECD. (2024). Rethinking health system performance assessment: A renewed framework. OECD Publishing.
42. Dixit, J., Prinja, S., Jyani, G., Bahuguna, P., Gupta, A., Vijayvergiya, R., & Kumar, R. (2023). Evaluating efficiency and equity of prevention and control strategies for rheumatic fever and rheumatic heart disease in India. Lancet Global Health, 11(3), e445–e455.
43. PATH. (2023). Private sector engagement for digitizing health care in India. PATH.
44. Medianama. (2023). NHA to set up 100 microsites across India to digitise healthcare data. Medianama.

Appendices

Acronyms and Glossary

1. **ABDM** – Ayushman Bharat Digital Mission: India's national digital health infrastructure, launched in 2021 under the National Health Authority (NHA).
2. **ABHA** – Ayushman Bharat Health Account: A unique digital health ID enabling longitudinal, consented access to personal health records.
3. **API** – Application Programming Interface: A set of open standards enabling secure and interoperable data exchange between digital-health applications.
4. **BCR** – Benefit-Cost Ratio: Ratio of discounted benefits to costs, used to assess investment efficiency.
5. **CERT-In / Health-CERT** – Computer Emergency Response Team / Sectoral CERT for Health: Agencies responsible for cybersecurity preparedness, monitoring, and incident response.
6. **DPI** - Digital Public Infrastructure
7. **DPDP Act 2023** – Digital Personal Data Protection Act, 2023: India's overarching legal framework for personal data protection.
8. **FHIR** – Fast Healthcare Interoperability Resources: The global HL7 data standard adopted by ABDM for structuring and exchanging electronic health information.
9. **HFR / HPR** – Health Facility Registry / Health Professional Registry: Foundational ABDM registries cataloguing verified facilities and practitioners.
10. **HTAIn** – Health Technology Assessment in India: The national framework guiding economic evaluation and cost-effectiveness analysis.
11. **ICER** – Incremental Cost-Effectiveness Ratio: A measure comparing additional costs per additional health outcome (e.g., ₹/QALY).
12. **IMDHC** – Inter-Ministerial Digital Health Council: Proposed statutory co-ordination body for ABDM governance across ministries and states
13. **λ_IN** – India-specific Affordability Threshold: Estimated at ₹14,000–₹22,000 per QALY, reflecting realistic public-sector willingness-to-pay
14. **NHA** – National Health Authority: The nodal agency responsible for ABDM and Ayushman Bharat PM-JAY.
15. **NHCX** – National Health Claims Exchange: The digital platform for standardized, paperless health claims processing
16. **NDHB 2019** – National Digital Health Blueprint: India's technical and governance blueprint for digital health.
17. **NPV** – Net Present Value: Present value of benefits minus costs, used to assess fiscal sustainability.
18. **PHR** – Personal Health Record: Citizen-controlled repository of health data linked through ABHA ID.
19. **QALY** – Quality-Adjusted Life Year: A health outcome measure combining quality and length of life gained.
20. **SROI** – Social Return on Investment: Monetary value of social and inclusion benefits generated by an investment.

21. **UHC** – Universal Health Coverage: The goal of equitable access to essential healthcare without financial hardship.
22. **WHO / OECD** – World Health Organization / Organisation for Economic Co-operation and Development: Global benchmark institutions for digital-health governance, evaluation, and comparators.

Detailed Methodology and Analytical Frameworks

Policy and Comparative Analysis

Framework: WHO Global Strategy on Digital Health (2020–2025) and OECD Digital-Health Readiness Framework were used as the guiding frameworks- these were supported with guiding questions for ABDM to be mapped to and aligned with specific regard to WHO strategic objectives and OECD surveys.

Core Pillar	Description	Guiding Questions	Mapping to International Frameworks
Strategy and Governance	Establishes the overarching vision, policy, and regulatory foundation for digital health. Ensures co-ordination across stakeholders, alignment with Universal Health Coverage (UHC) goals, and sustainability through legal, financial, and institutional mechanisms.	Is there a national digital health strategy with clear objectives? Does it align with UHC goals? Are governance and accountability structures defined across sectors?	WHO Strategic Objectives 1, 2, and 3
Infrastructure and Interoperability	Refers to the foundational digital and physical systems that enable data capture, storage, and exchange at scale. Includes electrification, registries, EHRs, connectivity, and adoption of common technical and semantic standards to ensure integration across health services.	Are core infrastructure elements (registries, EHRs, networks) available nationwide? Are interoperability standards (e.g.,HL7 FHIR) adopted and implemented across public and private systems?	WHO Strategic Objectives 2 and 3
Equity and Access	Ensures that digital health initiatives reduce disparities rather than exacerbate them. Focuses on inclusion across socioeconomic, geographic, and demographic groups through literacy, affordability, accessibility, and culturally appropriate tools.	Do digital health services reach underserved and rural populations? Are they affordable, multilingual, and designed for low-literacy groups? Are assistive technologies included?	WHO Strategic Objectives 2 and 4
Service Delivery and Innovation	Captures how digital tools are embedded into care delivery and health system functions. Emphasizes uptake, integration into clinical workflows, responsiveness to patient needs, and adaptability to emerging technologies.	Are digital tools like telemedicine, e-prescriptions, and AI-enabled tools widely adopted? Are health apps and digital platforms integrated into primary care? How is provider and patient engagement measured?	WHO Strategic Objectives 2 and 4
Monitoring and Impact	Focuses on continuous measurement of performance, accountability, and learning. Links investments in digital health to measurable improvement in efficiency, equity, and health outcomes.	Are there national evaluation frameworks and KPIs? Are value-for-money and cost-effectiveness analyses conducted? Are patient outcomes monitored and reported transparently?	WHO Strategic Objective 3 and M&E Action Plan

Dimension	Description	Guiding Questions	Mapping to International Frameworks
Analytic Readiness – Ability to access and link data (primary & secondary use)	Ability to access, link, and use data for healthcare (primary use) and for patient safety, public health, planning, system improvement, and research (secondary use).	Are national datasets available and interoperable? Do they cover the whole population? Can data be used for both care delivery and secondary purposes?	OECD Health Data Governance for the Digital Age (2022), Health at a Glance (pg 40–41, Fig 2.3, Table 4.5)
Analytic Readiness – Ability to access and link data (individual use)	Individuals have access to their own health records through secure portals and can interact with their data.	Do patients have secure, real-time access to their records? Can they share, manage, or correct information?	OECD 2021 Survey on EHRs, Health at a Glance (pg 42, Table 2.2)
Data Readiness – Governance of health data	Readiness to collect, store, and provide quality data through clear governance structures, defined accountability, and trusted policies.	Are governance frameworks published? Is there independent oversight, transparency, and trust-building mechanisms for citizens?	OECD Health Data Governance for the Digital Age (2022), Fig 2.4
Technology Readiness – Digital security	Secure connections, storage, and compliance with digital risk management frameworks for health data.	Are there national security protocols specific to health? Is there capacity to prevent and respond to cyber threats?	OECD Health at a Glance (2022), pg 47–48; Table 2.5
Technology Readiness – Certification of vendors	Certification process ensures vendors meet standards for messaging, terminology, and national EHR requirements.	Are there national certification frameworks? Do vendors comply with them? How widespread is adoption?	OECD 2021 Survey on EHR vendor certification, Health at a Glance (pg 49, Table 2.6)
Human Factor Readiness – Digital health strategies & governance	Strategies guided by equity, people-centeredness, integration, resilience, innovation, and workforce productivity.	Does the national digital health strategy align with WHO/OECD principles? Is there coherence across regions?	OECD Health at a Glance (2021), pg 50–51
Human Factor Readiness – Digital skills & health literacy	Population's ability to use digital tools and access online services effectively.	What share of the population has basic/advanced digital skills? Are there programmes to improve digital health literacy?	OECD Digital Health Literacy Country Reports
Human Factor Readiness – Citizen engagement	Ensuring people are central to health systems, with participation in design, monitoring, and evaluation.	Are citizens engaged in governance processes? Is there open data, feedback mechanisms, and transparency?	World Bank Digital Citizen Engagement Index, GTMI Dashboard; OECD Health at a Glance (2021, pg 53)

Process: Each policy source (National Health Policy 2017, National Health Stack 2018, National Digital Health Blueprint 2019, ABDM Building Block Study, NHA Annual Report 2022–23, ABDM Dashboard (August 2025), FHIR Implementation Guide 2024, SIDH/QCI/NABH certification materials, DHIS Guidelines, Digital Personal Data Protection Act 2023, and grievance and citizen-engagement mechanisms) was mapped to five pillars – Strategy & Governance; Infrastructure & Interoperability; Equity & Access; Service Delivery & Innovation; Monitoring & Impact - and four OECD dimensions - Human Factor Readiness, Technology Readiness, Analytic Readiness and Data Readiness.

Scoring: A five-point ordinal scale (1 = very limited → 5 = mature) was applied for 10 key documents to derive average pillar maturity.

Score	Integrated Scale - Policy, Planning & Readiness
1 - Very Limited	Fragmented or absent approach. Practices are informal, pilot-only, or scheme-specific. No national governance or strategy; limited awareness of global principles (e.g. OECD).
2 - Nascent	Early-stage national efforts exist but are partial or inconsistent. Policy or strategy documents may exist with limited operational traction. Coverage restricted to some states/programmes; weak or unenforced legal provisions.
3 - Developing	Formalized national frameworks and moderate implementation. Key registries or platforms operational in major schemes. Interoperability standards defined but not universal. Governance structures exist but adoption is uneven.
4 - Advanced	Broad adoption and integration across programmes, including private sector participation. Governance frameworks fully operational. Widespread use of interoperable digital IDs/EHRs. Demonstrated compliance with international benchmarks; controlled data access enabled for analytics.
5 - Mature	Institutionalized, continuously evaluated, and globally benchmarked. Fully interoperable national health data infrastructure used across public/private providers. Embedded digital literacy, strong enforcement, and open (governed) data access for research, policy, and innovation.

Mapping Simulation- Applying WHO Framework to ABDM: Building upon the customised structured scoring rubric metric (informed by WHO 2020-25 digital strategy objectives & strategies), this segment aims to integrate the scoring rubric, to map ABDM to the five identified pillars (i.e. strategy & governance, infrastructure & interoperability, equity & access, service delivery & innovation, and monitoring & impact) with the use of suggested guiding questions and the five-point scoring assessment.

Pillar	Metric	ABDM Score - Reasoning/Justification	Findings / Identified gaps (refer to 4.1 for more detail)
Strategy & Governance	National Strategy Alignment	**Score: 5** Presents clear objectives throughout digital health strategy by offering a government endorsed comprehensive, integrated national digital health strategy with alignment to UHC goals. NDH Blueprint emphasises the importance of keeping up with an evolving ecosystem, with principles and a 5-year timeline to guide the national strategy. (via NDH Blueprint pg 51-52, The Hindu - Cabinet Approval of ABDM Budget)Financial transparency and accountability are reflected. There is a INR 1600 crore budget allocated for this project between 2021-2026 - as of FY 2023-24, INR 586 crore (37%) has been allocated cumulatively (via Accountability Bharat Budget Links 2023).	N/A
	Governance Structure	**Score: 3** Governance and accountability structures exist across policies, however co-ordination between central and state parties is poorly defined and executed across both structures. State Governments are expected to play a dual role as a Government (providing support to Government of India for ABDM adoption, engagement with stakeholders) and as a Healthcare Provider (adopting digital health solutions, encouraging HPR/HFR registration, improving healthcare infrastructure) with guidance from NHA if required. District administrations are expected to engage with their communities and healthcare providers as well	Require feedback & accountability frameworks across layers of governance Need improved co-ordination frameworks between NHA & States, as well as among private stakeholders
Infrastructure & Interoperability	Core Digital Systems	**Score: 4** Developing core registry and healthcare network infrastructure nationwide is integral to this stage of ABDM - Digital Health Registries such as ABHA, HPR, HFR, Drug Registry are core components of Digital Public Goods within ABDM that are in the process of widespread uptake; they are live & nationally available with variation in adoption among states. To further reflect this emphasis on infrastructure building, ABDM's Technology Stack guidance provides structure to the core digital systems. The ABHA App provides a Personal Health Record, through linkage to ABHA-IDs, which can be accessed digitally. (via NHA Annual Report 2022-23)	Need further integration of digital systems across all tiers of healthcare system across all regions Require monitoring/t racking of infrastructure operability and user engagement
	Interoperability Standards	**Score: 4** Interoperability standards are being adopted across public & private systems however implementation across different regions & healthcare tiers are a challenge - ABDM follows the FHIR data standard for exchange and is implemented in clinical artifacts, billing artifacts, NHCX etc.The ABDM stack aims to create a digital public infrastructure which includes APIs (Telemedicine, Lab & Drugs, Bed Availability etc.) for health service. Data collected through the key registries (ABHA, HPR, & HFR) is integral to providing interoperability and free flow of secure health data between Health Information Providers & Health Information Users. (via Implementation Guide for Adoption of FHIR in ABDM and NHCX, ABDM Building Block Study)	Need for regulatory frameworks surrounding data collection and primary/ secondary use Need improved uptake of FHIR standards in real-life settings across rural and underserved areas; scaling up existing systems

Pillar	Metric	ABDM Score - Reasoning/Justification	Findings / Identified gaps (refer to 4.1 for more detail)
Equity & Access	Digital Inclusion	Score: 3Preliminary research reflects that digital health services don't yet consistently reach underserved and rural populations; this is heavily dependent on existing internet connectivity and pre-exisiting digital infrastructure of the region. Information available regarding assistive technologies remains limited. Although IEC activities are outlined by the NHA for ABDM (social media campaigns, online repository etc.) there does not appear to be an ABDM centred emphasis on digital training and literacy courses across providers and users available to bridge the digital divide. Important to note that with projects such as Indian National Optic Fiber Network still ongoing, there is not yet global internet access across India. There appears to be a lack of digital trust which proves to be a roadblock among users interacting on these public digital tools. (via Digital Literacy of Indian Health Workforce, NHA Annual Report 2022-23, Digital Foundations for Health Equity)	Minimal framework around improving digital infrastructure & digital literacy in rural India Lacking awareness of cultural and regional norms to be integrated to digital health initiatives and tool
	Affordability & Access	Score: 4ABDM places high value on ensuring that digital health services are affordable, multilingual and accessible at all levels of the public. However, tools do not appear necessarily aligned for low-literacy groups but there are capacity building activities aligned to improve global uptake. PM-JAY IT Systems are integrated with ABDM registries; the AB PM-JAY scheme ensures a large hospital network widely distributed geographically and provides access to affordable high-quality care. PHR apps are free to access, additionally digital tools and platforms being used by ANWs and ASHAs to better serve the public are also being leveraged to increase accessibility to high quality of healthcare, alongside access to telemedicine platforms such as eSanjeevani and AB HWC app. (via Digital Literacy of Indian Health Workforce, NHA Annual Report 2022-23, ABDM Building Block Study)	No assessment regarding reliability of PM-JAY interoperability with ABDM digital stack and data collection No data and planning to ensure accessibility of digital health tools among underserved areas No real-world focus on improving digital tool uptake in low-literacy groups
Service Delivery & Innovation	Integration into Care	Score: 3Digital tools via integrators and sandbox are an integral element of ABDM's technology stack; additionally digital public goods (i.e. UPI, Digilocker, e-RUPI), digital registries and health claims platforms are integrated within the technology stack. These allow clear and easy information flow between Health Information Providers (HIPs) and Users (HIUs). Digital health solutions can connect through APIs allowing users access to telemedicine services for instance. Measurement of engagement between HIP & HIUs requires further information and detailed evaluation. Integration of digital tools into primary care and global adoption is a core component of the technology stack, however, real-world challenges exist in barriers to adoption across rural or underdeveloped areas.	Digital tools not yet tailored to unique needs & requirements of different populations Poor adoption rates of tools among groups that are not digitally literate or lack access to internet
	Innovation Enablement	Score: 5ABDM has a very high capacity for innovation enablement; ABDM sandbox allows integration of current systems with building blocks and encourages participants' to innovate and develop new products, with a platform to test services that are registered. Digital Health Solution developers are encouraged to integrate with ABDM's API and DHIS incentivises this development as well; the ABDM sandbox is well structured with three milestones or stages of functionality that are defined. (via NHA Annual Report 2022-23, ABDM Building Block Study	N/A

Pillar	Metric	ABDM Score - Reasoning/Justification	Findings / Identified gaps (refer to 4.1 for more detail)
Equity & Access	Digital Inclusion	**Score: 3** Preliminary research reflects that digital health services don't yet consistently reach underserved and rural populations; this is heavily dependent on existing internet connectivity and pre-exisiting digital infrastructure of the region. Information available regarding assistive technologies remains limited. Although IEC activities are outlined by the NHA for ABDM (social media campaigns, online repository etc.) there does not appear to be an ABDM centred emphasis on digital training and literacy courses across providers and users available to bridge the digital divide. Important to note that with projects such as Indian National Optic Fiber Network still ongoing, there is not yet global internet access across India. There appears to be a lack of digital trust which proves to be a roadblock among users interacting on these public digital tools. (via Digital Literacy of Indian Health Workforce, NHA Annual Report 2022-23, Digital Foundations for Health Equity)	Minimal framework around improving digital infrastructure & digital literacy in rural India Lacking awareness of cultural and regional norms to be integrated to digital health initiatives and tool
	Affordability & Access	**Score: 4** ABDM places high value on ensuring that digital health services are affordable, multilingual and accessible at all levels of the public. However, tools do not appear necessarily aligned for low-literacy groups but there are capacity building activities aligned to improve global uptake. PM-JAY IT Systems are integrated with ABDM registries; the AB PM-JAY scheme ensures a large hospital network widely distributed geographically and provides access to affordable high-quality care. PHR apps are free to access, additionally digital tools and platforms being used by ANWs and ASHAs to better serve the public are also being leveraged to increase accessibility to high quality of healthcare, alongside access to telemedicine platforms such as eSanjeevani and AB HWC app. (via Digital Literacy of Indian Health Workforce, NHA Annual Report 2022-23, ABDM Building Block Study)	No assessment regarding reliability of PM-JAY interoperability with ABDM digital stack and data collection No data and planning to ensure accessibility of digital health tools among underserved areas No real-world focus on improving digital tool uptake in low-literacy groups
Service Delivery & Innovation	Integration into Care	**Score: 3** Digital tools via integrators and sandbox are an integral element of ABDM's technology stack; additionally digital public goods (i.e. UPI, Digilocker, e-RUPI), digital registries and health claims platforms are integrated within the technology stack. These allow clear and easy information flow between Health Information Providers (HIPs) and Users (HIUs). Digital health solutions can connect through APIs allowing users access to telemedicine services for instance. Measurement of engagement between HIP & HIUs requires further information and detailed evaluation. Integration of digital tools into primary care and global adoption is a core component of the technology stack, however, real-world challenges exist in barriers to adoption across rural or underdeveloped areas.	Digital tools not yet tailored to unique needs & requirements of different populations Poor adoption rates of tools among groups that are not digitally literate or lack access to internet
	Innovation Enablement	**Score: 5** ABDM has a very high capacity for innovation enablement; ABDM sandbox allows integration of current systems with building blocks and encourages participants' to innovate and develop new products, with a platform to test services that are registered. Digital Health Solution developers are encouraged to integrate with ABDM's API and DHIS incentivises this development as well; the ABDM sandbox is well structured with three milestones or stages of functionality that are defined. (via NHA Annual Report 2022-23, ABDM Building Block Study	N/A

Pillar	Metric	ABDM Score - Reasoning/Justification	Findings / Identified gaps (refer to 4.1 for more detail)
Monitoring & Impact	Evaluation Mechanisms	**Score: 2** Elements regarding evaluation of strategies exist but limited rigorous evaluation frameworks and defined KPIs present. Analyses regarding cost-effectiveness and value for money do not appear to be publicly available if present. Real-time monitoring regarding uptake of key building blocks publicly available and accessible via the ABDM dashboard is; these are mainly centered on digital public goods and registries, and additionally tracking microsite, DHIS and sandbox integrator progress. Capacity building exercises exist to address state-level challenges, grievance redressal portals.(via NHA Annual Report 2022-23, ABDM Dashboard)	Lack of rigorous evaluation frameworks and well-defined KPIs Minimal display of accountability & transparency to public through access to data regarding financial components and health outcomes No information regarding regular meetings with key stakeholders to receive feedback and progress of mission objectives
	Measurable Outcomes	**Score: 2** The overarching goal of ABDM is digitalisation of the healthcare system to improve healthcare delivery, access, and affordability. Digital solutions are encouraged, public digital tools are in place to improve interoperability and validity of providers, and the technology stack of ABDM is well defined. However, current evidence suggests that there does not appear to be much available data (quantitative or qualitative) provided by the government when it comes to tracking service quality or health outcomes linked to these digital tools. (via Unified Health Ecosystem in India, ABDM Assessment)	Poorly defined monitoring of progress to assess outcomes aligned with objectives (e.g. improving access to healthcare, ensuring equitable delivery of care, assessing efficiency of services) No available services to track engagement and outcome of digital tools among HIPs & HIUs

Mapping Simulation - Applying OECD Framework to ABDM: The four areas that are widely assessed and reviewed as part of OECD's digital health readiness dimensions are- analytic readiness, data readiness, technology readiness and human factor readiness. (14) These dimensions provide a framework to analyse dimensions of governance, interoperability, security, and infrastructure, through surveys and recommendations assessing Health Data Governance, Digital Security, Digital Identity and Digital Health Strategies. Using the four identified dimensions by the OECD digital health readiness, selected OECD described metrics to assess readiness in each dimension were chosen based on relevance to ABDM policies. Following that, the surveys used for evaluating progression in these metrics were mapped against ABDM responses and scored against an evaluation metric.

Dimension	Metric	ABDM Score and Reasoning/Justification	Gaps - Areas for improvement (refer to 4.1 for more detail)
Analytic Readiness	Ability to access and link data - primary & secondary use	Score: 2Several national datasets, relevant to OECD parameters, exist but coverage and breadth vary strongly. Few exist nationally and most offer incomplete population coverage. There is limited primary care clinical data and fragmented private-sector dataMost datasets are in the process of development and are still yet to see significant uptake across the population.Mechanism for dataset linkage and interoperability at a national scale is prioritised, however currently it does not meet OECD definitions.Further progress required regarding execution of transparent, published and audited dataset governance.(Sources - HMIS via https://www.hmis.mohfw.gov.in, Mental Health data via PMC: Mental health registry/data gaps, PM-JAY reports via PMC: Health data overview – India, Cancer Registry via PMC: Cancer registry coverage India, Mortality/Civil Registration System via https://crsorgi.gov.in, Patient experience & surveys via https://www.nfhsiips.in/nfhsuser/index.php, Population Census via Census of India , Heath Facility & Professional Registries via ABDM Facility Registry, ABDM Health Professional Registry (HPR))	Lack broad range of national datasets as required by OECD standardsPre-existing datasets often have incomplete population coverageUndefined primary and secondary use of existing data
	Ability to access and link data - individual use	Score: 3Patients will have real-time access to records and can share and manage information; patients can view and interact with their data through their ABHA-IDs linked to ABHA Health Records app.Patients can also link health records with other applications such as Arogya Setu; as per NRC for EHR standards, 'patients will have the sufficient privilege to inspect and view their medical records without any time limit.'Patient privileges to amend data shall be limited to correction of errors in the recorded patient/medical details. Patients remain skeptical regarding data security of their confidential and sensitive health information. (via https://phrbeta.abdm.gov.in/, https://www.nrces.in/standards/ehr-standards-for-india#strategic_highlights)	Minimal patient user engagement with health records to obtain feedback & improve servicesDistrust of e-services for health records
Data Readiness	Governance of health data	Score: 3India now has the Digital Personal Data Protection Act 2023, however, complete operationalism, audits and mapping to all 12 OECD principles described under this section are a work in progress. Uneven implementation of data quality, interoperability systems and standards across India and public/private sectors.Despite NHA security protocols and DPDP safeguard, there appear to be concerns regarding high vulnerability to data breaches in smaller states/hospitals. Absence of independent supervisory authority regarding governance, and no provisions for cross-border data sharing. Data linkage and accessibility is fragmented, secondary use is limited, there is no robust framework for linking data use to health outcomes. No large-scale national strategy appears to be developed for health workforce digital skills and capacity building, while health literacy for the population on the whole is limited. (via The Impact of the DPDP Act, 2023 on the Healthcare Industry: A Detailed Exploration, National Digital Health Blueprint (NDHB))	No airtight cybersecurity laws within healthcare sectorPoor integration of healthcare data regulatory frameworks into DPDP Act 2023Lack of uniform interoperability systems and standards across IndiaPoor data linkage across health systemsLack of defined accountable structures regarding health data protection

Dimension	Metric	ABDM Score and Reasoning/Justification	Gaps - Areas for improvement (refer to 4.1 for more detail)
Technology Readiness	Digital Security	Score: 3While legal foundations for digital security exist (DPDP 2023) and NHA has issued security protocols for data storage and encryption, health sector specific cybersecurity exercises are limited compared to OECD nations and smaller hospitals lack IT infrastructure and security. Furthermore no independent authority exists for health data breaches and there have been recent ransomware attacks (such as AIIMS 2022) that expose vulnerabilities in India's healthcare cybersecurity and technology readiness. Additionally, there appear to be key issues within the DPDP 2023 Bill as well which are contested among government officials and require further refinement following drafts starting from 2018. (via All India Institute of Medical Sciences (AIIMS), Delhi: Cyberattack Puts Digitalisation Under Scanner, Understanding India's New Data Protection Law | Carnegie Endowment for International Peace, https://prsindia.org/billtrack/digital-personal-data-protection-bill-2023)*Gap = Refer to 4.1	No independent authority for health data breachesNeed to prioritise widespread digital risk frameworks for health data Poor digital security in rural settings
	Certification of technology vendors in EHR systems	Score: 3National certification frameworks exist, however adoption is limited and compliance remains uneven.Vendors go through a certification process to comply with technical standards set by NHA under ABDM.Collaboration with partners (e.g. SIDH) allows seamless integration with ABDM for small to mid-sized hospitals, allowing them to get solutions certified and compliant, further enabling EMR and HMIS adoption . (via Progress on implementing & using EHR: OECD 2021, pg 73, Table D10)While the framework exists for vendors to integrate via APIs and certificate pathways are available, implementation is uneven.Many large hospitals and private chains pilot integration while district facilities and PHCs rely on fragmented or paper-based records. Additionally, many providers use local EMRs or software not yet ABDM compliant. (via System Integrators in Digital Health | SIDH Program by National Health Authority (NHA), NABH Draft Standards for HIS and EMR Systems July 2024, https://abdm.gov.in/qcicertified, NABH Certification Solutions, sandbox.abdm.gov.in, Full article: The Ayushman Bharat Digital Mission of India: An Assessment, NABH Releases Draft Standards for HIS & EMR Systems, Digital Health Standards for HIS/EMR Systems - NABH)	No uniform adoption and compliance of certification across vendorsLack of providers meeting ABDM requirements for national EHR systems

Dimension	Metric	ABDM Score and Reasoning/Justification	Gaps - Areas for improvement (refer to 4.1 for more detail)
Human Factor Readiness	Digital Health Strategies & Governance	Score: 4The three parameters where there seems to be a significant gap or lack of alignment, based on current review of literature, are as follows - improving productivity of health workforces, improving resilience and sustainability, and supporting learning health systems.ABDM adopts a clear national vision which is reiterated in the NHA and NDHM documents; this matches OECD/BMC recommendation. There is a clear definition of objectives, structure, funding and accountability which align with OECD standards. Key areas of alignment include: strong central leadership (NHA), technical standards & developer ecosystem, national registries, public reporting on KPI uptake (HFR, HPR, ABHA IDs, Sandbox integrators, Microsites) and ABDM linkage with health programs such as PM-JAY and eSanjeevani. Key gaps include lack of evaluation mechanisms measuring health system outcomes, no independent oversight/evaluation bodies, KPIs beyond infrastructure and registry uptake; furthermore state variation in uptake and readiness causes a lack of uniform commitment to objectives within the national vision. (via National eHealth strategies: a comparative study of nine OECD health systems , A brief guide on Ayushman Bharat Digital Mission (ABDM) and its various building blocks , A Unified Health Ecosystem ABDM, ABDM-Insights)	Require improve M&E frameworks to review cycles of ABDM and track progressNeed to narrow gap of ABDM alignment variation among StatesTracking of outcomes such as efficiency and productivity in heath settings using digital tools
	Digital skills of population & health literacy	Score: 2There is no direct comparison that can be made in line with this study, however pre-exisintg data established that only 38% of Indian households are digitally literate (further urban-rural divide of 61% in urban areas, 25% in rural areas, perpetrated by only 27% internet perpetration of rural areas). India ranks 73rd out of 120 countries for internet literacy as of 2021; additionally, there is no specific outline within ABDM to address digital literacy however schemes such as Digital India Program and PMGDISHA exist. (via Digital Literacy, Digital Literacy & Inclusion, India: internet literacy index by category 2021| Statista)	Lack of integrating digital literacy into ABDM parametersLack of data to assess digital literacy (e.g. census based questions)Need for improved internet literacy among populations and widespread connectivity to all regions
	Citizen Engagement & Public Involvement in Digital Health	Score: 2No available engagement in the Digital Citizen Engagement Index from the Indian Government - therefore it is difficult to accurately evaluate how engaged citizens are with governance processes.However, regarding general engagement and public involvement for ABDM, there appears to be capacity building exercises from stakeholders, training sessions for State Master Trainers and cross-learning workshops at the State Level. Formal citizen advisory tools, publicised consent tools and trust metrics remain limited. ABDM provides FAQs and dashboard metrics for public access, and a grievance redressal system open to citizens via a call-center that is available in 6 languages. (via GTMI World Bank Dashboard, NHA Annual Report 2022-23)	Minimal engagement with DCEI by GoIMinimal citizen engagement with ABDM through a lack of capacity building exercise and feedback loopsLack of publicly available information from grievance redressal system

Economic Analysis

The economic evaluation applied four complementary models- **ICER, NPV, BCR, and SROI-** in parallel to capture the **fiscal, clinical, and social dimensions** of value generated by the ABDM. Each model answers a different policy question and, when used together, provides a holistic view of cost-effectiveness, budget sustainability, and societal benefit.

Models Applied: ICER, NPV, BCR, and SROI were used in parallel to capture fiscal, clinical, and social dimensions of value.

Model	Definition	Policy Use
ICER **Incremental Cost-Effectiveness Ratio**	Measures the additional cost per incremental health benefit (e.g., ₹/QALY gained).	Determines whether an intervention is "worth it" relative to national willingness-to-pay.
NPV **Net Present Value**	Calculates discounted net fiscal benefit over time: future savings – current cost.	Tests long-term budget sustainability.
BCR **Benefit-Cost Ratio**	Ratio of total (discounted) benefits to total costs.	Communicates efficiency in simple terms for budget decisions.
SROI **Social Return on Investment**	Monetises non-financial outcomes such as equity, access, and time saved.	Integrates social value and inclusion into investment logic.

Reference Threshold: An India-specific affordability threshold (**λ_IN**) was derived to ground all ICER calculations in the real fiscal capacity of the public health system. Based on total public health expenditure of approximately ₹2.9 trillion (Union + State) and a population of ~1.4 billion, the per-capita spend equals ₹2,070 per year. Assuming one QALY represents one healthy life-year, and that roughly 0.7% of annual public expenditure could be reallocated to cost-effective innovations, the feasible affordability range becomes **₹14,000–₹22,000 per QALY**. This λ_IN represents India's budget-feasible marginal productivity threshold, distinct from aspirational GDP-based metrics, and forms the primary reference point for all cost-effectiveness analyses. The higher GDP-based threshold (₹2.1 lakh/QALY) was retained solely for sensitivity testing.

Assumptions: 5% annual discount rate, 10-year evaluation horizon, and conservative benefit realization lag of 3 years.

Sensitivity: Results stress-tested using GDP-based threshold (₹2.1 lakh/QALY) for upper-bound sensitivity.

Integration: The outputs of these models were mapped back to the five evaluation pillars - *Strategy & Governance, Infrastructure & Interoperability, Equity & Access, Service Delivery & Innovation, and Monitoring & Impact* - to identify where fiscal efficiency and social return are most pronounced.

Sentiment Analysis

Sample: The sentiment study was based on 20 semi-structured interviews conducted between March 2024 and July 2025 with a balanced representation of public, private, and civil-society stakeholders. The sample included six public-sector administrators (national and state level), five private hospitals and health-tech firms, four frontline providers (PHCs and district hospitals), and five representatives from insurers, TPAs, and civic-technology or patient-rights organisations.

Coding Framework: A directed content-analysis model was employed using six predefined codes drawn from the whitepaper's conceptual framework and WHO/OECD digital-readiness constructs:

- Trust_Infra: Confidence in ABDM's architecture, privacy, and data protection (Mean = +48).
- Friction_Adoption: Operational and workflow barriers to routine ABDM use (Mean = −67).
- Risk_Privacy: Concerns regarding data misuse, cybersecurity, and consent clarity (Mean = −32).
- Hope_Transform: Optimism about ABDM's transformative potential and national impact (Mean = +61).
- Apathy_Disengaged: Low motivation due to limited visible benefits or incentives (Mean = −41).
- Equity_Gap: Perceived inclusiveness gaps across socio-economic and linguistic divides (Mean = -54).

Validation: Each transcript was independently coded by two reviewers, and divergent interpretations were resolved through consensus. Findings were triangulated with secondary grey literature—including NASSCOM whitepapers, ABDM webinars, and press coverage—to minimise response bias and reinforce validity.

Outputs: The analysis produced mean sentiment scores, cross-code correlations, and a composite Net Sentiment Index (NSI) for each dimension. These results provide insight into behavioural barriers, levels of institutional trust, and perceived equity gaps, offering a qualitative complement to the policy and economic findings.

Triangulation and Integration

Findings from the **three methodological streams**—policy and benchmark review, economic evaluation, and stakeholder sentiment analysis—were **triangulated through a structured matrix-mapping approach**. Each pillar of the evaluation framework (Strategy & Governance, Infrastructure & Interoperability, Equity & Access, Service Delivery & Innovation, and Monitoring & Impact) was cross-referenced against the quantitative and qualitative evidence generated by these methods.

This enabled the **cross-validation of results,** identifying where technical readiness aligns with fiscal viability and public perception, and where discrepancies indicate systemic bottlenecks. The integrated insight layer, presented in Section 6, consolidates these findings into a coherent diagnostic of ABDM's performance maturity. It distils them into actionable reform levers for ABDM 2.0, focusing on institutional co-ordination, economic institutionalisation, inclusion, behavioural incentives, cyber trust, and outcome-oriented monitoring.

Category	Source Documents / Data Sets (with links)	Issuing Body / Year	Use in Analysis
Policy Frameworks	National Health Policy 2017 (PDF); National Health Stack (2018) (NITI Aayog brief in Annual Report, §3) (PDF); National Digital Health Blueprint 2019 (NDHB "comments invited" edition) (PDF); NDHM/ABDM Strategy Overview (PDF); ABDM launch/notifications (2021) (MoHFW press release) (page); DPDP Act, 2023 (Act PDF)	MoHFW; NITI Aayog; MeitY; NHA (2017–2023)	Foundational policy analysis (Ministry of Health and Family Welfare)
Operational Data	NHA Annual Reports (MoHFW DoHFW Annual Report includes NHA) (2023–24 PDF); National Health Accounts (Expenditure) (NHA 2021–22 page); ABDM Resources (Building blocks, handbook) (Resources hub); ABDM Sandbox (building blocks docs) (Sandbox)	MoHFW/NHA (2022–2025)	Quantitative performance benchmarking; build-out status (Ministry of Health and Family Welfare)
Technical Standards	ABDM FHIR Implementation Guide (current site, v6.5.0) (site); NRCeS FHIR IG (PDF) (PDF); Guide to Setup FHIR Terminology Server (PDF); NABH/QCI Digital Health certification (NABH Digital Health Certification Programme; QCI/NABH + ABDM accreditation announcement (PIB note)	NRCeS (2024–2025); QCI/NABH; NHA	Interoperability & compliance review (nrces.in)
Economic References	HTAIn Reference Case / Manual (2018) (PDF); Indian Reference Case (overview article) (Lancet SE Asia); National Health Accounts 2021–22 (records hub)	HTAIn/DHR; MoHFW/NHSRC (2018–2024)	λ_IN derivation; fiscal modelling inputs (htain.dhr.gov.in)
Comparative Benchmarks	WHO Global Strategy on Digital Health 2020–2025 (pub page); OECD Health at a Glance 2023 – Digital Health chapter (full report PDF); WHO/World Bank Global Digital Health Monitor (dashboard)	WHO; OECD; WHO/Partners (2020–2025)	Global policy & maturity comparison (World Health Organization)
Academic & Grey Literature	NASSCOM – Catalyzing Digital Health in India (2024) (page); NHA/ABDM webinar playlists (ABDM Webinar playlist; ABDM Sessions); Policy/industry briefs (e.g., NASSCOM DPI report) (PDF)	Industry & media; NHA (2023–2025)	Sentiment triangulation; ecosystem signals (NASSCOM Community)
Primary Qualitative Data	20 semi-structured interviews (Mar 2024–Jul 2025) – sampling frame and codebook retained in project files	Author fieldwork	Sentiment & perception analysis
Operational Schemes	Digital Health Incentive Scheme (DHIS) – Operational Guidelines (PDF)	NHA (2023)	Adoption incentives; uptake analysis (abdm.gov.in)
Grievance / Citizen-Engagement	NHA Grievance portal (via NHA site) (NHA site)	NHA (ongoing)	Governance, feedback loops (National Health Authority)
Analytical Tools	Excel models for ICER/NPV/BCR/SROI (author-developed); NVivo v14 project (author)	–	Quantitative & qualitative analysis integration

Publisher: Eliva Press Global Ltd

Email: info@elivapress.com

www.ingramcontent.com/pod-product-compliance
Lightning Source LLC
LaVergne TN
LVHW022322200926
848481LV00004B/1402
* 9 7 8 9 9 9 9 3 4 6 0 2 3 *